JUNIOR SCIENTISTS

Experiment with Weather

by Tamra B. Orr

CHERRY LAKE PUBLISHING • ANN ARBOR, MICHIGAN

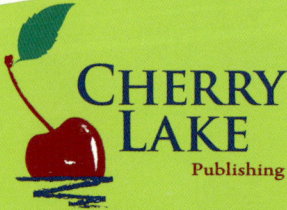

CHERRY LAKE
Publishing

Published in the United States of America by Cherry Lake Publishing
Ann Arbor, Michigan
www.cherrylakepublishing.com

Content Editor: Robert Wolffe, EdD, Professor of Teacher Education, Bradley University, Peoria, Illinois
Reading Adviser: Cecilia Minden-Cupp, PhD, Literacy Consultant

Design and Illustration: The Design Lab

Photo Credits: Page 11, ©Vasaleks/Shutterstock, Inc.; page 21, ©iStockphoto.com/quavondo; page 22, ©iStockphoto.com/hannerjo; page 27, ©iStockphoto.com/donald_gruener; page 28, ©iStockphoto.com/Fertnig; page 29, ©Ozerov Alexander/Shutterstock, Inc.

Library of Congress Cataloging-in-Publication Data
Orr, Tamra.
  Junior scientists. Experiment with weather / by Tamra B. Orr.
    p. cm.—(Science explorer junior)
  Includes bibliographical references and index.
  ISBN-13: 978-1-60279-841-0 (lib. bdg.)
  ISBN-10: 1-60279-841-9 (lib. bdg.)
  1. Meteorology—Experiments—Juvenile literature. 2. Science Projects—Juvenile literature. I. Title. II. Title: Experiment with weather. III. Series.
  QC863.5.O77 2010
  551.1078—dc22   2009048825

Portions of the text have previously appeared in *Super Cool Science Experiments: Weather* published by Cherry Lake Publishing.

Cherry Lake Publishing would like to acknowledge the work of The Partnership for 21st Century Skills. Please visit *www.21stcenturyskills.org* for more information.

Printed in the United States of America
Corporate Graphics Inc.
July 2010
CLFA07

Let's Experiment!

Science is fun!

Have you ever done a science **experiment**? They can be a lot of fun! You can use experiments to learn about almost anything.

4

TABLE OF CONTENTS

Scientists observe the world around them.

This book will help you learn how to think like a scientist. Scientists have a special way of learning new things. Some people call it the Scientific Method. This is how it often works:

- Scientists notice things. They **observe** the world around them. They ask questions about things they see, hear, taste, touch, or smell. They come up with problems they would like to solve.

A scientist guesses what the answer to her question is.

- They gather information. They use what they already know to guess the answers to their questions. This kind of guess is called a **hypothesis**.

- Then they test their ideas. They perform experiments or build models. They watch and write down what happens. They learn from each new test.

Scientists perform experiments to answer their questions.

- They think about what they learned and reach a **conclusion**. This means they come up with an answer to their question. Sometimes they **conclude** that they need to do more experiments!

Scientists like to come up with answers to their questions.

Conclusion

Are you ready to think like a scientist and learn about weather?

We will use the scientific method to learn more about weather. Where does wind come from? Why does it rain? We can try to answer our questions by doing experiments. Each experiment will teach us something new about weather. Are you ready to be a scientist?

Creating a Breeze

What kind of weather is this girl expecting?

What do we already know about weather? It is a big part of your life. It helps you choose what clothes to wear.

Weather is all about air, moisture, and **temperature**. Weather changes as air moves around. You may feel a warm breeze or a freezing blast of wind.

Why are some days hot and others cold?

What causes air to move around? Could it have anything to do with temperature? Let's do an experiment to find out. First, choose a hypothesis:

1. Temperature affects the way air moves.
2. Temperature does not affect the way air moves.

Let's get started!

Here's what you'll need:

- 2 metal cake pans of the same size. They should be labeled Pan #1 and Pan #2.
- Large cardboard box
- 2 heatproof pads of the same size
- Stick of incense
- Dry sand
- Ice
- Oven mitt
- Oven
- Timer
- Scissors
- Matches
- Adult helper

Gather everything you need before you begin.

Instructions:

1. Have an adult set the oven to the lowest heat setting.
2. Fill Pan #1 with 1 inch (2.5 centimeters) of sand. Use the oven mitt to place it in the oven.
3. Set the timer for 20 minutes.
4. Use the scissors to cut off the flaps of the cardboard box. Put the heatproof pads inside the box. The pads should sit side by side.

Make sure the pads are flat on the bottom of the box.

5. Fill Pan #2 with a 1-inch (2.5 cm) layer of ice.

6. Place Pan #2 on one of the pads.

7. Wait for the timer to go off. Use an oven mitt to pull Pan #1 from the oven. Place it on the other pad.

8. Ask an adult to use the matches to light the incense. Hold the incense stick so the burning tip is between the two pans. In which direction is the smoke blowing?

Can you see the smoke moving?

9. Write down what you notice.

Conclusion:

The smoke blew toward Pan #1. Why? The ice in Pan #2 cooled the air above it. This made the air sink. The hot sand in Pan #1 warmed the air. This warmer air rose. What happens as the warmer air rises? The cooler air moves in to take its place. This makes wind! Temperature does affect how air moves! Was your hypothesis correct?

Make It Rain!

Don't forget your umbrella on a rainy day!

Think of the last time it rained. Water that falls from the sky is called **precipitation**. What do you think causes precipitation?

16

We learned that air moves around when it mixes with different temperatures. Could temperature also have something to do with precipitation? Let's experiment! Choose a hypothesis:

1. Precipitation is caused by changes in temperature.
2. Precipitation is not caused by changes in temperature.

Let's get started!

Think carefully before choosing a hypothesis.

Here's what you'll need:

- Large glass jar with a wide opening
- Hot water
- Plate that is large enough to cover the mouth of the jar
- Ice cubes

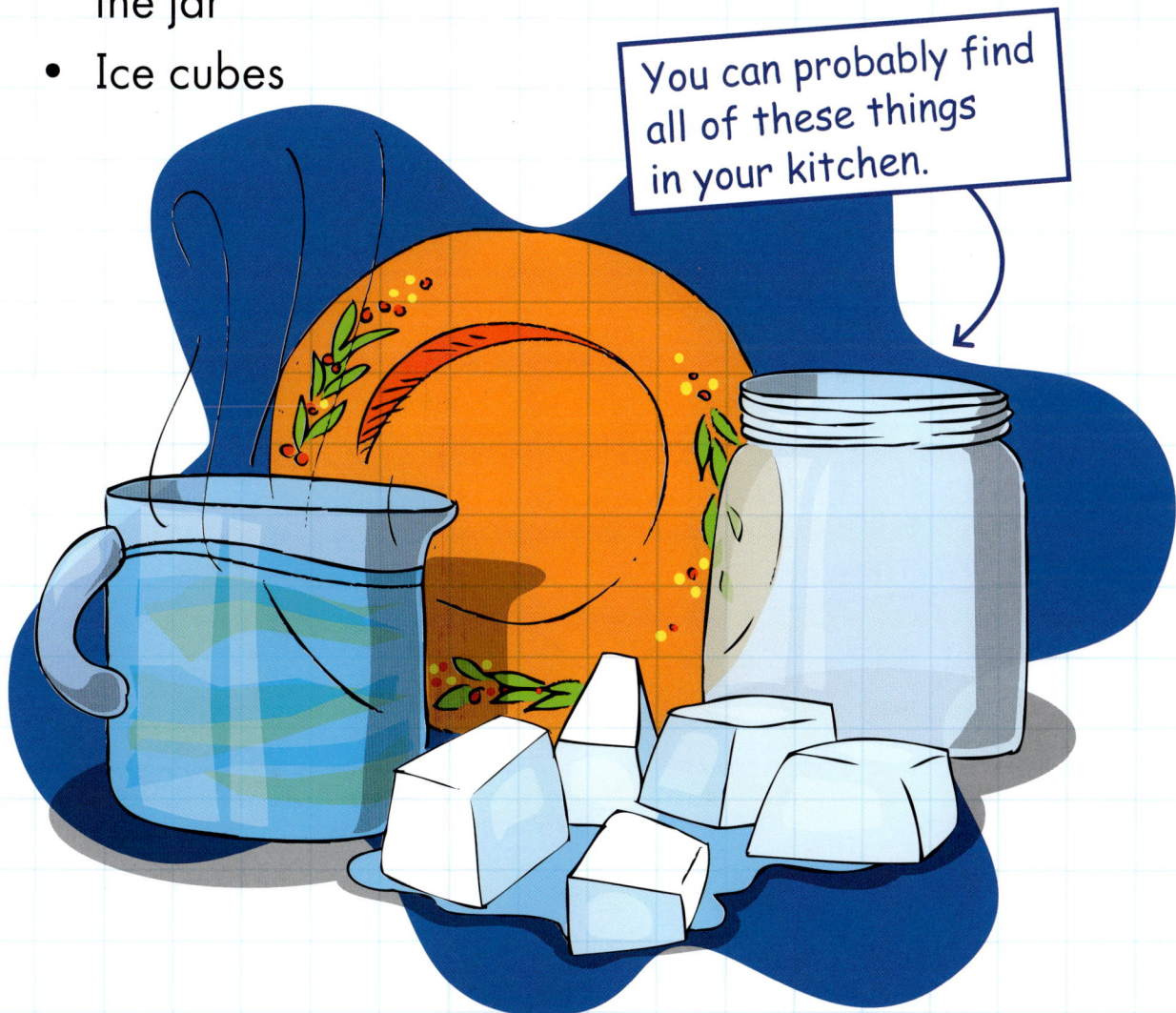

You can probably find all of these things in your kitchen.

Instructions:

1. Pour 2 inches (5 cm) of very hot tap water into the jar.

2. Cover the top of the jar with the plate. Let it sit for a few minutes. Can you see steam inside the jar? Does it remind you of a little cloud?

Observe the weather in your jar.

3. Place the ice cubes on the plate. Watch the bottom of the plate closely. What happens? Write down your observations.

What did you notice?

What did you learn about precipitation?

Conclusion:

Were water droplets forming at the bottom of the plate? The ice made the plate cold. The steam in the jar touched the cold plate. This formed drops of water.

This same thing happens up in the sky. Warm, moist air rises. It meets cold temperatures high up. Precipitation forms when the moisture cools. Was your hypothesis correct?

Nasty Weather

What causes storms?

Have you ever seen heavy rains in a summer storm? Do you ever wonder how they happen? Let's find out!

Storms form when warm air rises very quickly. Then cool air moves underneath. This makes precipitation fall from the clouds. Storms can dump heavy rains quickly. What causes the warm air to rise so fast? Could it have something to do with the cool air underneath? Time for another experiment!

We'll use ice and warm water to stand for the ideas of cold and warm air. How will the cold air (ice) and warm air (water) move around each other? Choose a hypothesis:

1. Cold water will flow above warm water.
2. Warm water will flow above cold water.

Let's get started!

Don't forget your notebook!

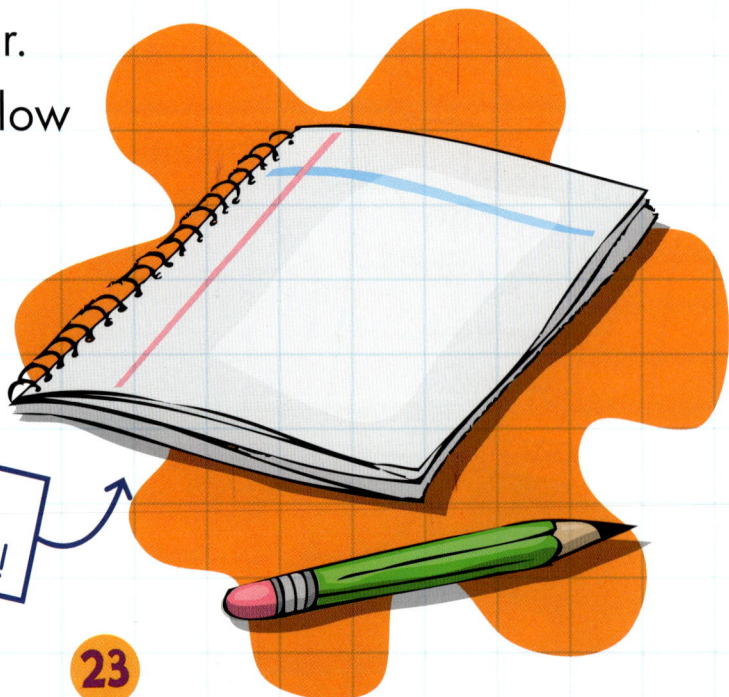

Here's what you'll need:

- Clear, plastic container about the size of a shoebox
- Warm water
- Blue ice cubes:
    - mix 5 to 6 drops of blue food coloring with 4 cups (1 liter) of water
    - pour into an ice cube tray
    - place the tray in a freezer
    - remove when the water is frozen
- 3 drops of red food coloring

You'll need to make the blue ice cubes before you begin the experiment.

Instructions:

1. Fill the container 2/3 full with warm water. Let it stand for 1 minute.

2. Put a blue ice cube at one end of the container.

3. Add 3 drops of red food coloring to the other end.

4. Watch carefully. Write down everything you see.

What happens when you add the drops of red food coloring?

Now you know more about how a big, windy storm starts!

Conclusion:

Think about the movement of the water. It stands for the movement of air in a storm. Did the blue water sink? This is because it was colder. It stands for a cold air mass. The red water rose. This is because it was warmer. It represents a warm air mass. Did you prove your hypothesis? Warm air can be forced to move up quickly when cold air moves toward it. You have modeled one part of how a storm forms!

Do It Yourself!

What kind of weather will you experiment with next?

Okay, scientists! Now you know more about weather. You learned how wind and precipitation are created. You also learned something about how a storm forms. You learned it all through your

observations and experiments. Try using the scientific method to answer other questions about weather.

A great time to spot rainbows is after rainstorms. Why? Could it have something to do with the moisture in the air? Come up with a hypothesis. Then ask an adult to help you create an experiment. Happy testing!

Do you think you can make a rainbow? Give it a try!

GLOSSARY

conclude (kuhn-KLOOD) to make a final decision based on what you know

conclusion (kuhn-KLOO-zhuhn) a final decision, thought, or opinion

experiment (ecks-PARE-uh-ment) a scientific way to test a guess about something

hypothesis (hy-POTH-uh-sihss) a guess about what will happen in an experiment

method (METH-uhd) a way of doing something

observe (uhb-ZURV) to see or notice something with the other senses

precipitation (pri-sih-puh-TAY-shuhn) rain, snow, sleet, or hail that falls toward the surface of Earth

temperature (TEM-pur-uh-chur) how hot or cold something is

FOR MORE INFORMATION

BOOKS

Pipe, Jim. *Earth's Weather and Climate*. Pleasantville, NY: Gareth Stevens Publishing, 2008.

Williams, Zella. *Experiments on the Weather*. New York: PowerKids Press, 2007.

WEB SITES

NASA—The Space Place: Make a Cloud Mobile
spaceplace.nasa.gov/en/kids/clouds/index.shtml
Try a fun craft project and learn more about clouds.

The Weather Channel—Kids!
www.theweatherchannelkids.com/
Learn about storms and more at this great weather site.

INDEX

ABOUT THE AUTHOR

Tamra B. Orr is a freelance author living in the Pacific Northwest. She has written more than 200 nonfiction books for readers of all ages. She and her four children, husband, cat, and dog all enjoy watching the region's weather. They experience tons of rain in the valley, oodles of snow in the mountains, hot temperatures in the desert, and lots of wind at the coast.